Sequential Fuzzy Clustering

A Sequential Bayesian Alternative to the Classical Parallel Fuzzy Clustering Model

Arash Abadpour

Arash Abadpour (arash@abadpour.com) is with the Imaging Group, Epson Edge, Epson Canada Limited.

The author acknowledges that a more concise version of this text is published in the journal of Information Sciences [1].

Contents

1 Introduction   1

2 Literature Review   3
    2.1 Notion of Membership   3
    2.2 Prototype-based Clustering   5
    2.3 Robustification   6
    2.4 Number of Clusters   8
    2.5 Weighted Clustering   9
    2.6 Sequential Clustering   9

3 Developed Method   11
    3.1 Model Preliminaries   11
    3.2 Clustering Model   14
    3.3 Assessment of Loss   15
    3.4 Alternating Optimization Strategy   17
    3.5 Outlier Rejection   21

4 Experimental Results   23
    4.1 Implementation Notes   24
    4.2 Grayscale Image Multi-Level Thresholding   24
    4.3 2-D Euclidean Clustering   27
    4.4 Plane Finding in Range Data   31

5 Conclusions   37

List of Figures

3.1 The classical clustering model utilized in FCM, PCM, and a significant number of other approaches in the literature. Here, the distances between a datum and a set of clusters are examined in parallel and a consolidation stage ensures that a sum-of-one constraint is satisfied. 14

3.2 Single cluster model utilized in the algorithm developed in this document. 15

4.1 Cluster and datum models in the grayscale image multi-level thresholding problem class. Here, a cluster is modeled as an interval on the grayscale axis, centered at ψ_c, and the datums are values on the grayscale axis. Each datum has a corresponding non-negative weight associated to it. The scale, λ, governs the size of the clusters. . 25

4.2 Results of the execution of the developed method for grayscale image multi-level thresholding. (a) Developed algorithm. (b) Classical FCM. 26

4.3 Results of the execution of the developed method for grayscale image multi-level thresholding. (a) Developed algorithm. (b) Classical FCM. 28

4.4 Cluster and datum models in the 2-D Euclidean clustering problem class. Here, a cluster is a modeled as a circle in the 2-D space, centered at ψ_c, and datums belong to a square in $\mathbb{R}^2$. Each datum has a corresponding non-negative weight associated to it. The scale, λ, governs the radius of the clusters. 29

4.5 Results of the execution of the developed method for 2-D Euclidean clustering. Size of the datums denote their weight and their shade of gray indicates the probability of being an inlier. (a) Developed algorithm. (b) Classical FCM. 30

4.6 Results of the execution of the developed method for 2-D Euclidean clustering. Size of the datums denote their weight and their shade of gray indicates the probability of being an inlier. (a) Developed algorithm. (b) Classical FCM. 31

4.7   Cluster and datum models in the plane finding in range data problem class. Here, a cluster is a modeled as a thick planar section of the 3-D space and is denoted by the closest point on it to the origin (more precise definition given in the text). Datums in this problem class are points in the 3-D space and distance between a datum and a cluster is defined as the distance between the corresponding point and the respective plane. The scale, λ, governs the thickness of the clusters.   33

4.8   Results of the execution of the developed method for plane finding in range data. (a) Developed algorithm. (b) Classical FCM.   34

4.9   Results of the execution of the developed method for plane finding in range data. (a) Developed algorithm. (b) Classical FCM.   35

5.1   The sequence-of-clusters model utilized in the algorithm developed in this work. Here, a sequence of C clusters process an input datum and produce C membership values and a probability value that indicates whether or not the datum is an outlier. Note that probability values are conditional.   50

Abstract

Unsupervised separation of a group of datums of a particular type, into clusters which are homogenous within a problem class-specific context, is a classical research problem which is still actively visited. Since the 1960's, the research community has converged into a class of clustering algorithms, which utilizes concepts such as fuzzy/probabilistic membership as well as possibilistic and credibilistic degrees. In spite of the differences in the formalizations and approaches to loss assessment in different algorithms, a significant majority of the works in the literature utilize the sum of datum-to-cluster distances for all datums and all clusters. In essence, this double summation is the basis on which additional features such as outlier rejection and robustification are built. In this work, we revisit this classical concept and suggest an alternative clustering model in which clusters function on datums sequentially. We exhibit that the notion of being an outlier emerges within the mathematical model developed in this document. Then, we provide a generic loss model in the new framework. In fact, this model is independent of any particular datum or cluster models and utilizes a robust loss function. An important aspect of this work is that the modeling is entirely based on a Bayesian inference framework and that we avoid any notion of engineering terms based on heuristics or intuitions. We then develop a solution strategy which functions within an Alternating Optimization pipeline.

Chapter 1

Introduction

Many signal and image processing applications require the unsupervised grouping of a set of datums of a particular model into a number of homogenous clusters. While the notion of homogeneity in these applications is defined within the context of the underlying physical phenomena, it is highly advantageous to be able to utilize abstract datum and cluster models and thus to arrive at a class-independent clustering algorithm.

In effect, any clustering algorithm is based on a number of models, some of which may seem trivial choices. Nevertheless, these structural elements dictate the behavior of the respective algorithms. For example, among the popular assumptions, within the field of data clustering, is the notion that datums can be reduced to multi-dimensional vectors and that clusters can be represented as prototypical datums. The Euclidean distance is a distance function of choice in such efforts. We argue that some of the major challenges in the field are direct results of these assumptions and therefore we suggest a critical perspective on *de facto* accepted aspects of popular data clustering algorithms.

Arguably, intuition and heuristics have been a driving force behind many major developments in the field of data clustering. Under such regimes, objective functions, constraints, and parameters are *engineered* based on a verbal description of certain perceptual notions, such as "tension", "force", and "seizure". This has led to different formalizations of the clustering problem which are often governed by regularization coefficients. As will be discussed in Chapter 2, these parameters often have to be set by a user and, generally, setting them improperly has severe implications for the performance of the algorithms. We argue in this document that a revisit of the construction methodology utilized by data clustering problems can alleviate this challenge.

In this document, we follow a Bayesian inference framework and derive a loss model for a generic clustering problem. The framework developed in this document replaces the well-known sum of datum-to-cluster distances with a sequential pipeline within which a notion of outlier rejection emerges. The model developed in this document is independent of the datum and cluster models applicable to any particular problem class and employs robustification means. We then provide a solution strategy for the developed clustering algorithm.

The rest of this document is organized as follows. First, in Chapter 2, we review the related literature and then, in Chapter 3, we present the developed method. Subsequently, in Chapter 4, we provide experimental results produced by the developed method on three different problem classes and, in Chapter 5, we present the concluding remarks.

Chapter 2

Literature Review

2.1 Notion of Membership

The notion of membership is a key point of distinction between different clustering schemes. Essentially, membership may be *Hard* or *Fuzzy*. Within the context of hard membership, each datum belongs to one cluster and is different from all other clusters. The fuzzy membership regime, however, maintains that each datum in fact belongs to all clusters, with the stipulation that the degree of membership to different clusters is different. K-means [2] and Hard C-means (HCM) [3] clustering algorithms, for example, utilize hard membership values. The reader is referred to [4] and the references therein for a history of K-means clustering and other methods closely related to it. Iterative Self-Organizing Data Clustering (ISODATA) [5] is a hard clustering algorithm as well.

With the introduction of Fuzzy Theory [6], many researchers incorporated this more natural notion into clustering algorithms [7]. The premise for employing a fuzzy clustering algorithm is that fuzzy membership is more applicable in practical settings, where generally no distinct line of separation between clusters is present. Additionally, from a practical perspective, it is observed that hard clustering techniques are extremely more prone to falling into local minima [8]. The reader is referred to [9] for the wide array of fuzzy clustering methods developed in the past few decades.

Initial work on fuzzy clustering was done by Ruspini [10] and Dunn [11] and it was then generalized by Bezdek [9] into Fuzzy C-means (FCM). In FCM, datums, which are denoted as $x_1, \cdots, x_N$, belong to $\mathbb{R}^k$ and clusters, which are identified as $\psi_1, \cdots, \psi_C$, are represented as points in $\mathbb{R}^k$. FCM makes the assumption that the number of clusters, C, is known through a separate process or expert opinion and minimizes the following objective function,

$$\Delta = \sum_{c=1}^{C} \sum_{n=1}^{N} f_{nc}^m \|x_n - \psi_c\|^2. \qquad (2.1)$$

This objective function is heuristically suggested to result in appropriate clustering results and is constrained by,

$$\sum_{c=1}^{C} f_{nc} = 1, \forall n. \qquad (2.2)$$

Here, $f_{nc} \in [0, 1]$ denotes the membership of datum x_n to cluster ψ_c.

In (2.1), $m > 1$ is the *fuzzifier* (also called *weighing exponent* and *fuzziness*). The optimal choice for the value of the fuzzifier is a debated matter [12] and is suggested to be "an open question" [13]. Bezdek [14] suggests that $1 < m < 5$ is a proper range and utilizes $m = 2$. The use of $m = 2$ is suggested by Dunn [11] in his early work on the topic as well and also by Frigui and Krishnapuram [15], among others [16]. Bezdek [17] provided physical evidence for the choice of $m = 2$ and Pal and Bezdek [18] suggested that the best choice for m is probably in the interval $[1.5, 2.5]$. Yu, Cheng, and Huang [13] argue that the choices for the value of m are mainly empirical and lack a theoretical basis. They worked on providing such a basis and suggested that "a proper m depends on the data set itself" [13].

Recently, Zhou, Fu and Yang [19] proposed a method for determining the optimal value of m in the context of FCM. They employed four Cluster Validity Index (CVI) models and utilized repeated clustering for $m \in [1.1, 5]$ on four synthetic data sets as well as four real data sets adopted from the UCI Machine Learning Repository [20] (refer to [21] for a review of CVIs and [22] for coverage in the context of relational clustering). The range for m in that work is based on previous research [23] which provided lower and upper bounds on m. The investigation carried in [19] yields that $m = 2.5$ and $m = 3$ are optimal in many cases and that $m = 2$ may in fact not be appropriate for an arbitrary set of datums. This result is in line with other works which demonstrate that larger values of m provide more robustness against noise and the outliers. Nevertheless, significantly large values of m are known to push the convergence towards the sample mean, in the context of Euclidean clustering [13]. Wu [24] analyzes FCM and some of its variants in the context of robustness and recommends $m = 4$.

Rousseeuw, Trauwaert and Kaufman [25] suggested replacing f_{nc}^m with $\alpha f_{nc} + (1-\alpha) f_{nc}^2$, for a known $0 < \alpha < 1$. Klawonn and Hoppner suggested to generalize this effort and to replace f_{nc}^m with an increasing and differentiable function $g(f_{nc})$ [26, 27].

Pedrycz [28, 29, 30] suggested to modify (2.2) in favor of customized $\sum f_{nc}$ constraints for different values of n. That technique allows for the inclusion of *a priori* information into the clustering scheme and is addressed as Conditional Fuzzy C-means (CFCM). The same modification is carried out in Credibilistic Fuzzy C-Means (CFCM) [31], where, the "credibility" of datums is defined based on the distance between datums and clusters. Therefore, in that approach, (2.2) is modified in order to deflate the membership of outliers to the set of clusters (also see [32]). Customization of (2.2) is also carried out in Cluster Size Insensitive FCM (csiFCM) [33] in order to moderate the impact of datums in larger clusters on an smaller adjacent cluster. Leski [12] provides a generalized version of this approach in which $\sum \beta f_{nc}^{\alpha}$ is constrained.

2.2 Prototype-based Clustering

It is a common assumption that the notion of homogeneity depends on datum-to-datum distances. This assumption is made implicitly when clusters are modeled as *prototypical* datums, also called *clustroids* or cluster *centroids*, as in FCM, for example. A prominent choice in these works is the use of the Euclidean distance function [34]. For example, the potential function approach considers datums as energy sources scattered in a multi-dimensional space and seeks peak values in the field [35] (also see [36, 37]). We argue, however, that the *distance* between the datums may not be either defined or meaningful and that what the clustering algorithm is to accomplish is the minimization of *datum-to-cluster* distances. For example, when datums are to be clustered into certain lower-dimensional subspaces, as is the case with Fuzzy C-Varieties (FCV) [38], the Euclidean distance between the datums is irrelevant.

Nevertheless, prototype-based clustering does not necessarily require explicitly present prototypes. For example, in kernel-based clustering, it is assumed that a non-Euclidean distance can be defined between any two datums. The clustering algorithm then functions based on an FCM-style objective function and produces clusteroids which are defined in the same feature space as the datums [39]. These cluster prototypes may not be explicitly represented in the datum space, but, nevertheless, they share the same mathematical model as the datums [40] (the reader is referred to a review of Kernel FCM (KFCM) and Multiple-Kernel FCM (MKFCM) in [41] and several variants of KFCM in [42]). Another example for an intrinsically prototype-based clustering approach in which the prototypes are not explicitly "visible" is the Fuzzy PCA-guided Robust k-means (FPR k-means) clustering algorithm [43] in which a centroid-less formulation [44] is adopted which, nevertheless, defines homogeneity as datum-to-datum proximity. Relational clustering approaches constitute another class of algorithms which are intrinsically based on datum-to-datum distances (for example refer to Relational FCM (RFCM) [45] and its non-Euclidean extension Nerf C-means [46]). The goal of this class of algorithms is to group datums into *self-similar* bunches. Another algorithm in which the presence of prototypes may be less evident is Multiple Prototype Fuzzy Clustering Model (FCMP) [47], in which datums are described as a linear combination of a set of prototypes, which are, nevertheless, members of the same $\mathbb{R}^k$ as the datums are. Additionally, some researchers utilize L_r-norms, for $r \neq 2$ [48, 49, 50, 51], or other datum-to-datum distance functions [52].

We argue that a successful departure from the assumption of prototypical clustering is achieved when clusters and datums have different mathematical models. For example, the Gustafson-Kessel algorithm [53] models a cluster as a pair of a point and a covariance matrix and utilizes the Mahalanobis distance between datums and clusters (also see the Gath-Geva algorithm [54]). Fuzzy shell clustering algorithms [16] utilize more generic geometrical structures. For example, the FCV [38] algorithm can detect lines, planes, and other hyper-planar forms, the Fuzzy C Ellipsoidal Shells (FCES) [55] algorithm searches for ellipses, ellipsoids, and hyperellipsoids, and the Fuzzy C Quadric Shells (FCQS) [16] and its variants seek quadric and hyperquadric clusters (also see Fuzzy C Plano-Quadric Shells (FCPQS) [56]).

2.3 Robustification

Dave and Krishnapuram [57] argue that the concept of membership function in FCM and the concept of weight function in robust statistics are related. Based on this perspective, it is argued that the classical FCM in fact provides an indirect means for attempting robustness. Nevertheless, it is known that FCM and other least square methods are highly sensitive to noise [31]. Hence, there has been ongoing research on the possible modifications of FCM in order to provide a (more) robust clustering algorithm [58, 59]. Dave and Krishnapuram [57] provide an extensive list of relevant works and outline the intrinsic similarities within a unified view (also see [60, 61]).

The first attempt to robustifying FCM, based on one account [57], is the Ohashi Algorithm [60, 62]. That work adds a noise class to FCM and writes the robustified objective function as,

$$\Delta = \alpha \sum_{c=1}^{C} \sum_{n=1}^{N} f_{nc}^m \|x_n - \psi_c\|^2 + (1-\alpha) \sum_{n=1}^{N} \left(1 - \sum_{c=1}^{C} f_{nc}\right)^m. \tag{2.3}$$

The transformation from (2.1) to (2.3) was suggested independently by Dave [61, 63] when he developed the Noise Clustering (NC) algorithm as well. The core idea in NC is that there exists one additional imaginary prototype which is at a fixed distance from all datums and represents noise. That approach is similar to modeling approaches which perform consecutive identification and deletion of one cluster at a time [64, 65]. Those methods, however, are expensive to carry out and require reliable cluster validity measures.

Krishnapuram and Keller [66] extended the idea behind NC and developed the Possibilistic C-means (PCM) algorithm by rewriting the objective function as,

$$\Delta = \sum_{c=1}^{C} \sum_{n=1}^{N} t_{nc}^m \|x_n - \psi_c\|^2 + \sum_{c=1}^{C} \eta_c \sum_{n=1}^{N} (1 - t_{nc})^m. \quad (2.4)$$

Here, t_{nc} denotes the degree of representativeness or *typicality* of datum x_n to cluster ψ_c (also addressed as *possibilistic degree* in contrast to the *probabilistic* model utilized in FCM). As expected from the modification in the way t_{nc} is defined, compared to that of f_{nc}, PCM removes the sum of one constraint, shown in (2.2), and in effect extends the idea of one noise class in NC into C noise classes. In other words, PCM could be considered as the parallel execution of C independent NC algorithms, each seeking a cluster. Therefore, the value of C is somewhat arbitrary in PCM [57]. For this reason, PCM has been called a *mode-seeking* algorithm where C is the upper bound on the number of modes.

We argue that the interlocking mechanism present in FCM, i.e. (2.2), is valuable in that, not only clusters seek homogenous sets, but that they are also forced into more optimal "positions" through forces applied by competing clusters. In other words, borrowing the language used in [34], in FCM clusters "seize" datums and it is disadvantageous for multiple clusters to claim high membership to the same datum. There is no phenomenon, however, in NC and PCM which corresponds to this internal factor. Additionally, it is likely that PCM clusters coincide and/or leave out portions of the data unclustered [67]. In fact, it is argued that the fact that at least some of the clusters generated through PCM are nonidentical is because PCM gets trapped into local minimum [68]. PCM is also known to be more sensitive to initialization than other algorithms in its class [34].

It has been argued that both concepts of possibilistic degrees and membership values have positive contributions to the purpose of clustering [69]. Hence, Pal, Pal, and Bezdek [70] combined FCM and PCM and rewrote the optimization function of Fuzzy Possiblistic C-Means (FPCM) as minimizing,

$$\Delta = \sum_{c=1}^{C} \sum_{n=1}^{N} (f_{nc}^m + t_{nc}^\eta) \|x_n - \psi_c\|^2, \quad (2.5)$$

subject to (2.2) and $\sum_{n=1}^{N} t_{nc} = 1, \forall c$. That approach was later shown to suffer from different scales for f_{nc} and t_{nc} values, especially when $N \gg C$, and, therefore, additional linear coefficients and a PCM-style term were introduced to the objective function [71] (also see [72] for another variant). It has been argued that the resulting objective function employs four correlated parameters and that the optimal choice for them for a particular problem instance may not be trivial [34]. Additionally, in the new combined form, f_{nc} cannot necessarily be interpreted as a membership value [34].

Weight modeling is an alternative robustification technique and is exemplified in the algorithm developed by Keller [73], where the objective function is rewritten as,

$$\Delta = \sum_{c=1}^{C} \sum_{n=1}^{N} f_{nc}^{m} u_c \frac{1}{\omega_n^q} \|x_n - \psi_c\|^2, \qquad (2.6)$$

subject to $\sum_{n=1}^{N} \omega_n = \omega$. Here, the values of ω_n are updated during the process as well.

Frigui and Krishnapuram [15] included a robust loss function in the objective function of FCM and developed Robust C-Prototypes (RCP),

$$\Delta = \sum_{c=1}^{C} \sum_{n=1}^{N} f_{nc}^{m} u_c \left(\|x_n - \psi_c\| \right). \qquad (2.7)$$

Here, $u_c(\cdot)$ is the robust loss function for cluster c. They further extended RCP and developed an unsupervised version of RCP, nicknamed URCP [15]. Wu and Yang [40] used $u_c(x) = 1 - e^{-\beta x^2}$ and developed Alternative HCM (AHCM) and Alternative FCM (AFCM) algorithms (also see [74]).

2.4 Number of Clusters

The classical FCM and PCM, and many of their variants, are based on the assumption that the number of clusters is known (the reader is referred to [9, Chapter 4] and [75, Chapter 4] for reviews of this topic). While PCM-style formulations may appear to relax this requirement, the corresponding modification is carried out at the cost of yielding an ill-posed optimization problem [34]. Hence, provisions have been added to existing clustering algorithms in order to address this challenge. Repeating the clustering procedure for different numbers of clusters [54, 76, 77] and Progressive Clustering are two of the approaches devised in the literature.

Among the many variants of Progressive Clustering are methods which start with a significantly large number of clusters and freeze "good" clusters [76, 78, 79, 56], approaches which combine compatible clusters [80, 76, 78, 79, 56, 15], and the technique of searching for one "good" cluster at a time until no more is found [64, 79, 81]. Use of regularization terms in order to push the clustering results towards the "appropriate" number of clusters is another approach taken in the literature [82]. These regularization terms, however, generally involve additional parameters which are to be set carefully, and potentially per problem instance (for example see the mixed C-means clustering model proposed in [70]).

Dave and Krishnapuram conclude in their 1997 paper that the solution to the general problem of robust clustering when the number of clusters is unknown is "elusive" and that the techniques available in the literature each have their limitations [57]. In this document, we acknowledge that the problem of determining the appropriate number of clusters is hard to solve and even hard to formalize. Additionally, we argue that this challenge is equally applicable to many clustering problems independent of the particular clustering model utilized in the algorithms. Therefore, we designate this challenge as being outside the scope of this contribution and assume that either the appropriate number of clusters is known or that an exogenous means of cluster pruning is available which can be utilized within the context of the algorithm developed in this document.

2.5 Weighted Clustering

Many fuzzy and possibilistic clustering algorithms make the assumption that the datums are equally important. Weighted fuzzy clustering, however, works on inputs datums which have an associated positive weight [64]. This notion can be considered as a marginal case of clustering fuzzy data [83]. Examples for this setting include clustering of a weighted set, clustering of sampled data, clustering in the presence of multiple classes of datums with different priorities [84], and a measure used in order to speed up the execution through data reduction [85, 86, 87]. Nock and Nielsen [88] formalize the case in which weights are manipulated in order to move the clustering results towards datums which are harder to include regularly. Note that the extension of FCM on weighted sets has been developed under different names, including Density-Weighted FCM (WFCM) [86], Fuzzy Weighted C-means (FWCM) [89], and New Weighted FCM (NW-FCM) [90].

2.6 Sequential Clustering

FCM, PCM, and the majority of other available clustering techniques, expose all datums to all clusters simultaneously. This is in effect the results of loss models such as (2.1), in which NC datum-to-cluster distance terms are combined directly. We argue, however, that there are advantages in constructing a sequential clustering framework.

A first approach to restructuring the operation of the clustering algorithm into a sequence, is the technique of detecting one cluster at a time and then removing the corresponding datums from the set. More elaborate works in this category avoid deletion of clustered datums and utilize terms which disincentivize the consideration of previously clustered datums for new clusters. For example, in [91], the authors detect the clusters sequentially and then utilize a regularization term which pushes new clusters off the territory of previous ones. The employed regularization term, however, depends on a set of β_i values and the authors state that "[s]ince we do not have an established method for settings β_is, these parameters must be set empirically". As discussed before, we argue that the dependence of the algorithm on such configuration parameters is a significant practical challenge for the use of an algorithm in the context of unsupervised operation.

A second class of sequential clustering approaches serializes the datums and allows the clusters to respond to them in a sequence. Fuzzy Competitive Learning (FCL) [92] and its predecessor Enhanced Sequential Fuzzy Clustering (ESFC) [93] are sample algorithms in this category. This approach is either adopted as a measure to avoid the entrapment of the clusters in local minima, or, in response to the characteristics of problems classes in which datums are in fact discovered sequentially and are not available to the algorithm as a full set.

Chapter 3

Developed Method

In this section, we develop a novel alternative to the generic clustering model utilized in FCM, PCM, and a significant number of other fuzzy, possibilistic, and credibilistic algorithms available in the literature. We show that the hegemony of the classical model, which calculates the sum total of datum-to-cluster distances, can be broken by a novel clustering model which also provides better robustification in addition to an emergent outlier rejection property.

We start this chapter with presenting model preliminaries in Section 3.1 and then we develop a novel clustering model in Section 3.2. The assessment of loss in this model, as outlined in Section 3.3, results in the formalization of the fuzzy clustering problem as an optimization problem, for which a solution strategy, based on Alternative Optimization, is proposed in Section 3.4. Finally, we discuss the emergence of a notion of outlier rejection within the framework of the developed model in Section 3.5.

3.1 Model Preliminaries

We assume that a problem class is given, within the context of which a datum model is known and denote a datum as x. We also assume that a particular cluster model is provided, which complies with the *notion of homogeneity* relevant to the problem class at hand, and denote a cluster as ψ.

In this work, we utilize a weighted set of datums, defined as,

$$\mathbf{X} = \left\{ (\omega_n; x_n) \right\}, n = 1, \cdots, N, \omega_n > 0, \tag{3.1}$$

and we define the *weight* of **X** as,

$$\Omega = \sum_{n=1}^{N} \omega_n. \tag{3.2}$$

In this context, when estimating loss, we treat **X** as a set of realizations of the random variable x and write,

$$p\{x_n\} = \frac{\omega_n}{\Omega}. \tag{3.3}$$

We assume that the real-valued positive *distance function* $\phi(x, \psi)$ is defined. Through this abstraction, we allow for any applicable distance function and therefore avoid the dependence of the underlying algorithm on the Euclidean or other specific notions of distance. As examples, when the datums belong to $\mathbb{R}^k$, the Euclidean Distance, any L_r norm, and the Mahalanobis Distance are special cases of the notion of datum-to-cluster distance defined here. The corresponding cluster models in these cases denote $\psi \in \mathbb{R}^k$, $\psi \in \mathbb{R}^k$, and ψ identifying a pair of a member of $\mathbb{R}^k$ and a $k \times k$ covariance matrix, respectively.

We assume that $\phi(x, \psi)$ is differentiable in terms of ψ and that for any non-empty weighted set **X**, the following function of ψ,

$$\Delta_{\mathbf{X}}(\psi) = E\{\phi(x, \psi)\} = \frac{1}{\Omega} \sum_{n=1}^{N} \omega_n \phi(x_n, \psi), \tag{3.4}$$

has one and only one minimizer which is also the only solution to the following equation,

$$\sum_{n=1}^{N} \omega_n \frac{\partial}{\partial \psi} \phi(x_n, \psi) = 0, \tag{3.5}$$

In this document, we assume that a function $\Psi(\cdot)$ is given, which, for the input weighted set **X**, produces the optimal cluster ψ which minimizes (3.4) and is the only solution to (3.5). We address $\Psi(\cdot)$ as the *cluster fitting function*. In fact, $\Psi(\cdot)$ is the solution to the M-estimator given in (3.4). Examples for $\Psi(\cdot)$ include mean and median when x and ψ are real values and,

$$\phi(x, \psi) = (x - \psi)^2, \tag{3.6}$$

and,

$$\phi(x, \psi) = |x - \psi|, \tag{3.7}$$

respectively. We note that when a closed-form representation for $\Psi(\cdot)$ is not available, conversion to a W-estimator can produce a procedural solution to (3.5) (refer to [94, 95] for details).

We assume that a function $\Psi_\circ(\cdot)$, which may depend on $\mathbf{X}$, is given, that produces an appropriate number of initial clusters. We address this function as the *cluster initialization function* and denote the number of clusters produced by it as C. In this work, as discussed in Section 2.4, we make the assumption that the management of the value of C is carried out by an external process which may also intervene between subsequent iterations of the algorithm developed in this document.

We assume that a robust loss function, $u(\cdot) : [0, \infty] \to [0, 1]$, is given. Here, the assumption is that,

$$\lim_{\tau \to \infty} u(\tau) = 1. \tag{3.8}$$

We assume that $u(\cdot)$ is an increasing differentiable function which satisfies $u(0) = 0$ and $u(\lambda) = \frac{1}{2}$, for a known value of $\lambda > 0$, which we address as the *scale* parameter (note the similarity with the cluster-specific weights in PCM [66]). In fact, λ has a similar role to that of *scale* in robust statistics [96] (also called the *resolution* parameter [36]) and the idea of distance to noise prototype in NC [61, 63]. Scale can also be considered as the controller of the boundary between inliers and outliers [57]. From a geometrical perspective, λ controls the radius of spherical clusters and the thickness of planar and shell clusters [34]. One may investigate the possibility of generalizing this unique scale factor into cluster-specific scale factors, i.e. λ_c values, in line with η_c variables in PCM [66].

In this work, we utilize the rational robust loss function given below,

$$u(x) = \frac{x}{\lambda + x}. \tag{3.9}$$

Hence, we model the loss of the datum x_n when it belongs to cluster c as,

$$u_{nc} = u\Big[\phi(x_n, \psi_c)\Big]. \tag{3.10}$$

Note that varying λ does not impact the overall geometrical layout of $u(x)$. In fact, λ linearly stretches the vertical span of the function. To exhibit this point, we make λ explicit in the notation of $u(x)$ and write $u_\lambda(\lambda x) = u_1(x)$.

We acknowledge that one may consider the possibility of utilizing Tukey's biweight [97], Hampel [98], and Andrews loss functions in the present framework. Huber and Cauchy loss functions are not bounded and therefore are not applicable to this work. Refer to [57, Table I] for mathematical formulations and to [99] for an implementation which utilizes Huber within the context of FCM.

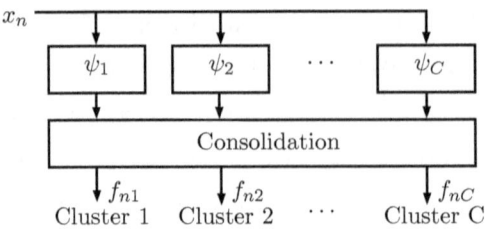

Figure 3.1: The classical clustering model utilized in FCM, PCM, and a significant number of other approaches in the literature. Here, the distances between a datum and a set of clusters are examined in parallel and a consolidation stage ensures that a sum-of-one constraint is satisfied.

3.2 Clustering Model

As discussed in Section 3.1, a typical clustering algorithm in the literature employs a notion of datum-to-cluster distance which relates an arbitrary datum to an arbitrary cluster. The next step is then to generalize this relationship between an arbitrary datum and a set of clusters. In this framework, the dominant model utilized in the literature is that the clusters "compete" for a datum simultaneously, thus resulting in a weighted-sum-of-distances aggregation model. In effect, this model is utilized in FCM and PCM and the whole spectrum of the approaches which combine and modify these two algorithms in different forms.

As shown in Figure 3.1, the classical model is based on a set of clusters which each issues a verdict for any datum and a subsequent consolidation stage which ensures that a sum-of-one constraint, generally in the form shown in (2.2), is satisfied. A key challenge with this model, however, is that it assumes that every datum is an inlier. While, as discussed in Section 2.3, further developments utilize more complex models in order to address this aspect, these approaches often depend on entities, including model variables and regularization weights, which are hard to perceptually define and challenging to assign a value to. Moreover, these extensions to the classical models are often based on engineered terms and heuristics-driven processes which are later justified through experimental results.

We argue that a core challenge in the classical models is that they utilize clusters as independent points of datum assessment. In other words, in these models, each cluster provides a level of interest in a datum without respect for the relationship between the other clusters and the same datum. In effect, the consolidation stage is required in those models in order to avoid the challenges present in the original PCM model, as discussed in Section 2.3. To our understanding, more efficient execution

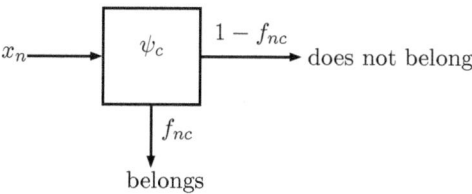

Figure 3.2: Single cluster model utilized in the algorithm developed in this document.

of cluster-to-cluster "communication" is an avenue which can explored.

Therefore, we propose the cluster model exhibited in Figure 3.2. In this model, a cluster "observes" a datum and makes a decision as of the probability that the datum belongs to the cluster. Note that, in this model, the process of decision-making, which in the classical model is postponed until after the consolidation stage, is also carried out directly by the cluster. We show in this document that this cluster model allows for the serialization of the clustering process and that it has important benefits at an affordable cost.

Utilizing the two-output cluster model exhibited in Figure 3.2, we construct the sequence of clusters shown in Figure 5.1. Here, clusters function within a Bayesian structure, where cluster c "notices" datum x_n, after it is observed by clusters 1 to $c-1$.

3.3 Assessment of Loss

We first discuss the membership probabilities corresponding to the clusters in the model depicted in Figure 5.1 for the arbitrary datum x_n. In order to formalize these probabilities, we define the C sets $\tilde{\mathbf{X}}_c, c = 1 \cdots C$, where $\tilde{\mathbf{X}}_c$ is the set of datums which belong to cluster c when a Maximum Likelihood selection process is carried out. We define the set $\mathbf{X} - \tilde{\mathbf{X}}$ as the set of outliers. Here,

$$\tilde{\mathbf{X}} = \bigcup_{c=1}^{C} \tilde{\mathbf{X}}_c. \qquad (3.11)$$

As seen in Figure 5.1, the first cluster decides to "choose" x_n independent of the decisions made by the other clusters. Hence,

$$p\left\{x_n \in \tilde{\mathbf{X}}_1\right\} = f_{n1}. \qquad (3.12)$$

However, when $c > 1$, we need to include the complete set of probabilities that lead to x_n being passed on to cluster c in the derivations. Hence, for $c = 2$ we write,

$$p\left\{x_n \in \tilde{\mathbf{X}}_2\right\} = f_{n2}(1 - f_{n1}), \qquad (3.13)$$

and for $c = 3$ we have,

$$p\left\{x_n \in \tilde{\mathbf{X}}_3\right\} = f_{n3}(1 - f_{n2})(1 - f_{n1}), \tag{3.14}$$

Similarly,

$$p\left\{x_n \in \tilde{\mathbf{X}}_c\right\} = f_{nc} \prod_{c'=1}^{c-1} (1 - f_{nc'}). \tag{3.15}$$

Therefore, the probability that x_n is an outlier is calculated as,

$$p\left\{x_n \in \mathbf{X} - \tilde{\mathbf{X}}\right\} = 1 - \sum_{c=1}^{C} p\left\{x_n \in \tilde{\mathbf{X}}_c\right\} = \tag{3.16}$$

$$1 - \sum_{c=1}^{C} \left[f_{nc} \prod_{c'=1}^{c-1} (1 - f_{nc'}) \right] = \prod_{c'=1}^{C} (1 - f_{nc'}).$$

There are multiple methods for proving the last equality in (3.16). One may do the derivation through defining,

$$F_{c_1,c_2} = 1 - \sum_{c=c_1}^{c_2} \left[f_{nc} \prod_{c'=c_1}^{c-1} (1 - f_{nc'}) \right], \tag{3.17}$$

and subsequently utilizing reverse induction in order to start from $F_{C,C} = 1 - f_{nC}$ and to calculate $F_{1,C}$.

We use these results in order to assess the aggregate loss in the clustering model shown in Figure 5.1 corresponding to the arbitrary datum x_n. We first write,

$$E\left\{\text{Loss}\middle|x_n\right\} = \sum_{c=1}^{C} E\left\{\text{Loss}\middle|x_n \in \tilde{\mathbf{X}}_c\right\} p\left\{x_n \in \tilde{\mathbf{X}}_c\right\} \tag{3.18}$$

$$+ E\left\{\text{Loss}\middle|x_n \in \mathbf{X} - \tilde{\mathbf{X}}\right\} p\left\{x_n \in \mathbf{X} - \tilde{\mathbf{X}}\right\}.$$

Then, we use (3.15) and (3.16) and rewrite (3.18) as,

$$E\left\{\text{Loss}\middle|x_n\right\} = \sum_{c=1}^{C} \left[u_{nc} f_{nc} \prod_{c'=1}^{c-1} (1 - f_{nc'}) \right] + \prod_{c'=1}^{C} (1 - f_{nc'}). \tag{3.19}$$

Here, the loss of an outlier is modeled as 1, the value to which the robust loss function converges to when $\phi(x_n, \psi_c)$ is very large.

Now, we use (3.3) and combine the single-datum loss model derived in (3.19) for all datums in order to estimate the aggregate loss in the system as,

$$E\{\text{Loss}|\mathbf{X}\} = \frac{1}{\Omega}\sum_{n=1}^{N}\omega_n\left[\sum_{c=1}^{C}\left[u_{nc}f_{nc}\prod_{c'=1}^{c-1}(1-f_{nc'})\right]\right] \qquad (3.20)$$
$$+\frac{1}{\Omega}\sum_{n=1}^{N}\omega_n\prod_{c'=1}^{C}(1-f_{nc'}).$$

Hence, we construct the clustering algorithm developed in this document as an optimization problem in which the following objective function is to be minimized,

$$\Delta = \sum_{n=1}^{N}\sum_{c=1}^{C}\omega_n\left[f_{nc}\prod_{c'=1}^{c-1}(1-f_{nc'})\right]u_{nc} + C\frac{1}{C}\sum_{n=1}^{N}\omega_n\prod_{c'=1}^{C}(1-f_{nc'}). \qquad (3.21)$$

Here, we have dropped the constant Ω^{-1} factor and we have chosen to rewrite the constant factor 1 as CC^{-1} for reasons which will become clear later.

Note that, in (3.21), the only requirement for the the value of f_{nc} is that $f_{nc} \in [0,1]$. In other words, a sum-of-one constraint, similar to the one given in (2.2), is not required here. The relaxation of this requirement in the method developed in this document, however, is distinctly different from the approach in PCM, as in this work the clusters are inherently tied together. We emphasize that not only the disadvantageous separation of clusters, as observed in the original PCM, is not present in this work, but also that a concept of outlier rejection emerges within the model developed in this document. Additionally, as will be shown later, the condition $f_{nc} \in [0,1]$ will be organically satisfied in this work.

Close assessment of (3.21) shows that this cost function complies with an HCM-style hard template. It is known, however, that migration from HCM towards a fuzzy/probabilistic formulation has important benefits, as outlined in Section 2.1. Hence, defining $U = C^{1-m}$, we rewrite the aggregate loss as,

$$\Delta = \sum_{n=1}^{N}\sum_{c=1}^{C}\omega_n\left[f_{nc}^m\prod_{c'=1}^{c-1}(1-f_{nc'})^m\right]u_{nc} + U\sum_{n=1}^{N}\omega_n\prod_{c'=1}^{C}(1-f_{nc'})^m. \qquad (3.22)$$

3.4 Alternating Optimization Strategy

In this section, we provide a solution strategy for minimizing (3.22). This process utilizes a set of input clusters and carries a variant of the well-known Alternating Optimization (AO) methodology. This process is composed of a number of subsequent stages. In each step, the objective function is minimized relative to a subset of the clustering unknowns. In fact, the procedure developed in this work contains two stages of assignment and reclustering. During the assignment stage, the cluster representations are assumed to be known and the optimal values of f_{nc} are estimated. The reclustering stage, on the other hand, fits optimal clusters to the set of datums given their known membership status to the clusters. Here, these two processes are described.

The classical approach to utilizing AO in the context of fuzzy clustering is to work on the partial derivative of Δ relative to f_{nc} while (2.2) is incorporated into the objective function using Lagrange Multipliers. However, because the structure of (3.22) is different from the classical objective functions present in the fuzzy/possibilistic literature, a different approach is required in the present work. Additionally, (2.2) is not applicable to the present work.

We define,

$$\Delta_{c^*}(n) = \sum_{c=c^*}^{C} \left[f_{nc}^m \prod_{c'=c^*}^{c-1} (1 - f_{nc'})^m \right] u_{nc} + U \prod_{c'=c^*}^{C} (1 - f_{nc'})^m. \qquad (3.23)$$

One may in fact show that $\Delta_{C+1}(n) = U$ and, using (3.22), that,

$$\Delta = \sum_{n=1}^{N} \omega_n \Delta_1(n). \qquad (3.24)$$

Now, direct derivation shows that,

$$\Delta_{c^*}(n) = f_{nc^*}^m u_{nc^*} + (1 - f_{nc^*})^m \Delta_{c^*+1}(n). \qquad (3.25)$$

Note that an important result of (3.25) is that $\Delta_{c^*+1}(n)$ is independent of f_{nc^*}.

Now, we address the process of minimizing $\Delta(n)$, i.e. contribution of x_n to Δ, through finding the optimal values for f_{nc} for $c = 1, \cdots, C$. We have,

$$\frac{\partial}{\partial f_{nc^*}} \Delta(n) = \frac{\partial}{\partial f_{nc^*}} \Delta_1(n) = \frac{\partial}{\partial f_{nc^*}} \left(f_{n1}^m u_{n1} + (1 - f_{n1})^m \Delta_2(n) \right) \qquad (3.26)$$

$$= (1 - f_{n1})^m \frac{\partial}{\partial f_{nc^*}} \Delta_2(n).$$

Continuing this derivation proves that,

$$\frac{\partial}{\partial f_{nc^*}} \Delta(n) = \prod_{c'=1}^{c^*-1} (1 - f_{nc'})^m \frac{\partial}{\partial f_{nc^*}} \Delta_{c^*}(n) = \qquad (3.27)$$

$$\prod_{c'=1}^{c^*-1} (1 - f_{nc'})^m \left[m f_{nc^*}^{m-1} u_{nc^*} - m(1 - f_{nc^*})^{m-1} \Delta_{c^*+1}(n) \right].$$

Equating (3.27) with zero yields,

$$f_{nc^*} = \frac{\Delta_{c^*+1}(n)^{\hat{m}}}{u_{nc^*}^{\hat{m}} + \Delta_{c^*+1}(n)^{\hat{m}}}. \tag{3.28}$$

Here, and throughout the derivations, we use the following notation in order to simplify the formulations,

$$\hat{m} = \frac{1}{m-1}. \tag{3.29}$$

Now, we plug the value of f_{nc^*} from (3.28) in (3.25) and produce the following simplified update equation,

$$\Delta_{c^*}(n) = (1 - f_{nc})^{m-1} \Delta_{c^*+1}(n). \tag{3.30}$$

Hence, we propose the procedure shown in Algorithm 1 for calculating f_{nc}.

> **for** $n = 1$ *to* N **do**
> $\quad \Delta = C^{1-m};$
> $\quad$ **for** $c = C$ *down to* 1 **do**
> $\quad\quad f_{nc} = \frac{\Delta^{\hat{m}}}{u_{nc}^{\hat{m}} + \Delta^{\hat{m}}};$
> $\quad\quad \Delta = (1 - f_{nc})^{m-1} \Delta;$
> $\quad$ **end**
> **end**

Algorithm 1: The procedure for calculating f_{nc}.

Note that, as discussed in Section 2.1, there is general consensus in the community that $m = 2$ is an appropriate choice for the value of the fuzzifier. Setting $m = 2$ results in $m - 1 = \hat{m} = 1$, which simplifies many of the formulations given in this document. Due to shortage of space we do not carry these simplified equations. Note that one may utilize the Klawonn-Hoppner generalization [26] of the fuzzifier within the context of the algorithm developed in this document. Also, one can directly show that this process always guarantees that $0 \leq f_{nc} \leq 1$, as required by the model.

As discussed in Section 2.1, there is general consensus that $m = 2$ is an appropriate choice for the value of fuzzifier. Hence, we note that when $m = 2$, $m - 1 = \hat{m} = 1$ and therefore the update procedure for f_{nc} is simplified to the following,

Now, we develop the update procedure for the clusters. This stage is similar to the classical

for $n = 1$ *to* N **do**
$\quad \Delta = \frac{1}{C}$;
$\quad$ **for** $c = C$ *down to* 1 **do**
$\quad\quad f_{nc} = \frac{\Delta}{u_{nc} + \Delta}$;
$\quad\quad \Delta = (1 - f_{nc})\Delta$;
$\quad$ **end**
end

FCM. We write,

$$\frac{\partial}{\partial \psi_c}\Delta = \sum_{n=1}^{N} \omega_n \left[f_{nc}^m \prod_{c'=1}^{c-1} (1 - f_{nc'})^m \right] u'_{nc} \frac{\partial}{\partial \psi} \phi(x_n, \psi) \quad (3.31)$$

Here, we have used the substitute notation,

$$u'_{nc} = \frac{d}{d\tau} u \bigg|_{\tau = \phi(x_n, \psi_c)}. \quad (3.32)$$

Now, using (3.5) we have,

$$\psi_c = \Psi\left(\left\{ \omega_n f_{nc}^m u'_{nc} \prod_{c'=1}^{c-1} (1 - f_{nc'})^m, x_n \right\}\right). \quad (3.33)$$

Note, however, that the closed form given in (3.33) is based on ignoring the dependence of u'_{nc} on ψ_c. Hence, we propose to utilize (3.33) in order to produce the new cluster representation $\psi_c^{\star}$. Then, we only use this cluster representation if replacing the current ψ_c with $\psi_c^{\star}$ results in a reduction in the value of Δ.

An alternative method is to utilize the technique developed by Weiszfeld [100] and Miehle [101, 102] (similar techniques are cited under different names as well [103, 104]). The Weiszfeld technique utilizes the fixed point method in order to solve (3.5) when $\phi(\cdot)$ is not the Euclidean distance function (refer to [103] for details and to [105] for options for the acceleration of the technique). A weighted version of the Levenberg-Marquardt algorithm [106] may also be applicable for certain distance functions and loss functions.

Here, a note on the rational robust loss function, given in (3.34), is necessary. In previous works [107], we utilized the quadratic robust loss function, defined as,

$$u(x) = \frac{x^2}{\pi \lambda \left(\lambda^2 + x^2\right)}. \quad (3.34)$$

However, we have empirically noticed that the use of the rational function results in higher reclustering rates, compared to what is achieved when the quadratic function is utilized. Thus, we adopt the rational function in this work. Investigation of the possible causes of this phenomenon and the potential for finding a robust loss function which would maximize the reclustering rate is outside the scope of this work.

3.5 Outlier Rejection

For certain classes of problems, it is not a requirement that every datum must in fact be assigned to a cluster. In other words, the problem class under consideration may prefer a solution in which clusters are not unnecessarily "bloated" in order to include outliers. We satisfy the requirements of such problem classes through returning C disjoint sets $\tilde{\mathbf{X}}_c, c = 1, \cdots, C$, as the inliers, as well as $\mathbf{X} - \tilde{\mathbf{X}}$ as the set of outliers. This approach has precedence in the literature (for example refer to the use of the reject class in [108]). We emphasize that the classical model used in FCM and a number of its variants is based on the assumption that every datum is an inlier.

An interesting observation about the model developed in Section 3.2 is the emergence of the notion of the probability that a datum is an outlier. In effect, the model developed in this document defines f_{nc} variables as conditional probabilities of belonging to the clusters. However, through the process, the probability of belonging to none of the clusters emerges, which we derive an analytic form for in (3.16).

We propose to utilize a Maximum Likelihood inference framework and to compare the probability values corresponding to $x_n \in \tilde{\mathbf{X}}_c$, for $c = 1, \cdots, C$ as well as the probability that $x_n \in \mathbf{X} - \tilde{\mathbf{X}}$. Derivation shows that a datum belongs to one of the $\tilde{\mathbf{X}}_c$, and therefore is an inlier, when,

$$\max_c \left\{ f_{nc} \prod_{c'=1}^{c-1} (1 - f_{nc'}) \right\} > \prod_{c'=1}^{C} (1 - f_{nc'}). \tag{3.35}$$

Here, we have used (3.15) and (3.16).

Hence, if the problem is to be solved in *inclusive* mode, then datum x_n is assigned to cluster c_n where,

$$c_n = \arg\max_c \left\{ f_{nc} \prod_{c'=1}^{c-1} (1 - f_{nc'}) \right\} \tag{3.36}$$

In *non-inclusive* mode, however, c_n is not defined if x_n is not an inlier, i.e. if it does not satisfy (3.35). This strategy has similarities to Conditional Fuzzy C-means (CFCM) [12], but the method developed in this document does not require the *a priori* knowledge necessary in CFCM.

Chapter 4

Experimental Results

A common practice in many clustering works is to provide experimental results as evidence for the superiority of the methods developed therein. We argue that this after-the-fact evaluation approach is a direct implication of the fact that many of these works utilize objective functions and/or constraints which are selected based on intuition. In other words, our argument is that the methodology utilized when the clustering problem is formalized must be validated as well. Hence, whether a particular approach is capable of producing acceptable results for a certain number of problems is a weak indicator of its usability. We find the necessity of validating the construction methodology behind a clustering problem to be important from an epistemological perspective.

As presented in Chapter 3, this work constructs the formalization of the clustering problem through Bayesian inference. In this process, we actively avoid the use of our intuition in generating variables, the objective function, and constraints, and strictly require the formalization to comply with an explicitly stated mathematical model. This is in direct contrast to FCM, PCM, and a majority of other methods in the field, which despite their magnificent achievements, are in essence formalizations based on what the respective designers of the algorithms have found to be appropriate.

Nevertheless, in this section, we provide sample experimental results produced through the application of the method developed in this document on multiple problem instances within the context of three problem classes. In each case, we provide comparisons with the outcomes of the classical FCM.

Note that the two algorithms are always executed using identical input datums and there is no difference in parameters or strategies utilized by the two algorithms. In effect, the comparison provided in this section is at the level of objective functions. Nevertheless, we emphasize that this presentation is by no means intended to be a complete evaluation process. As such, the contribution of this document is a novel modeling framework and the emergent properties of that. Hence, the carried results are to be treated as samples utilized in order to exhibit the differences of the previous model and the presented one.

First, in Section 4.1, a few notes on the implementation of the developed algorithm are given. Then, in Sections 4.2, 4.3, and 4.4, the deployment of the developed algorithm for three different problem classes are discussed.

4.1 Implementation Notes

The developed algorithm is implemented as a class named *Belinda* in MATLAB Version 8.1 (R2013a). It takes use of Image Processing Toolbox for minor image-related primitives and the major operations are implemented as C/MEX dll's. The code is executed on a Personal Computer which runs Windows 7, 64bit, on an Intel Core i5-2400 CPU, 3.10GHz, with 8.00GB of RAM.

Each problem class is implemented as a child class for *Belinda*. The child classes implement a constructor which creates the weighted set $\mathbf{X}$ based on the input image, data file, etc. The child classes also implement the three functions $\phi(\cdot)$, $\Psi_\circ(\cdot)$, and $\Psi(\cdot)$ and set the value of λ independent of $\mathbf{X}$. The child classes are not responsible for any of the core operations of the developed algorithm. These operations are implemented in the parent class *Belinda*. This class abstracts datums and clusters and operates on them indifferent to the content of the problem class-specific entities.

4.2 Grayscale Image Multi-Level Thresholding

The problem of grayscale image multi-level thresholding defines datums as grayscale values and models a cluster as an interval on the grayscale axis centered at the scalar ψ_c [64]. Figure 4.1 provides a visual representation of cluster and datum models in this problem.

In order to produce the datums, we calculate the histogram of the input image (32 bins). Here, each bin represents one datum and the number of pixels in the bin, normalized over the area of the image, produces the corresponding weight. Distance between a datum and a cluster is defined as the square difference and the initial clusters are defined as uniformly-distributed points in the working range. The cluster fitting function in this case calculates the weighted sum of the input elements. This problem class utilizes a scale of 25 gray levels and is defined as inclusive.

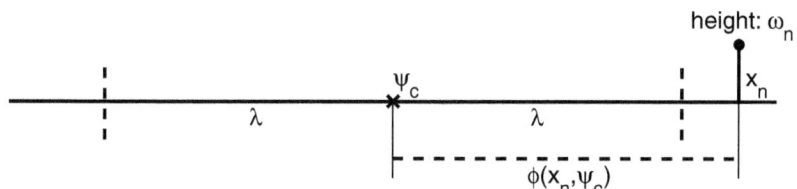

Figure 4.1: Cluster and datum models in the grayscale image multi-level thresholding problem class. Here, a cluster is modeled as an interval on the grayscale axis, centered at ψ_c, and the datums are values on the grayscale axis. Each datum has a corresponding non-negative weight associated to it. The scale, λ, governs the size of the clusters.

Figure 4.2 presents the results of running the developed algorithm on the standard image *Cameraman* alongside the results produced by the classical FCM for the same sample image. Here, in visualization, x_n is replaced with ψ_{c_n} in order to exhibit the clusters. We note that meaningful comparison of the two outputs in the image domain is challenging. Therefore, we review the layout of the clusters in the histogram domain.

The two graphs given in Figure 4.2 visualize the histogram of the input image overlaid by the converged clusters as well as the probabilities that the datums are inliers. In these visualizations, the thick dashed line denotes the histogram of the input image, thus effectively visualizing the values of ω_n. The thin dashed line exhibits the possibility of being an inlier for x_n, which, as expected, is always one for FCM. The colored curves in these figures denote membership to the converged clusters.

We first note that FCM clusters are bulkier than those produced by the proposed algorithm, as demonstrated by wider range of high membership values in Figure 4.2-(b)-bottom compared to Figure 4.2-(a)-bottom. Therefore, we argue that the proposed method produces clusters which are more focused on a homogenous subset of the input datums. A side achievement of this phenomenon is that FCM clusters claim high membership values to outliers. This is for example visible in the right end of Figure 4.2-(b)-bottom, where the curve representing the third cluster resides on a significant height. We compare this situation to the low level of membership claimed within the context of the developed algorithm to the same set of datums (see the right end of Figure 4.2-(a)-bottom).

This result also indicates the difference in the computational complexity of the developed

Figure 4.2: Results of the execution of the developed method for grayscale image multi-level thresholding. (a) Developed algorithm. (b) Classical FCM.

method compared to the classical FCM. In fact, the developed method takes more iterations to converge (27 for the developed algorithm vs. 11 for classical FCM) and also takes more time to settle into the final solution (100 milliseconds for the developed algorithm vs. 17 milliseconds for classical FCM). This is due to the more complicated nature of the calculation of f_{nc} in the developed algorithm as well as the appropriateness check required after each call to $\Psi\left(\cdot\right)$ in this algorithm. In fact, in an experiment which involved the execution of the two algorithms on 212 standard images, adopted from [109] and USC-SIPI, the developed algorithm required 98 ± 57 milliseconds to complete while the classical FCM converged in 35 ± 16 milliseconds. Thus, the developed algorithm is in this case about 3.15 times more expensive than the classical FCM. The input images in this experiment contain 512×512 pixels.

Figure 4.3 shows the output generated by the developed algorithm compared to that of the classical FCM for the standard image *Lena*. This comparison validates the same observations made in Figure 4.2, i.e. the clusters generated by the developed algorithm are more specific and that FCM clusters claim high membership to outliers.

4.3 2-D Euclidean Clustering

The problem of 2-D Euclidean Clustering is the classical case of finding dense clusters in weighted 2-D data. Here, dense is defined in terms of inter-cluster proximity of points. Figure 4.4 provides a visual representation of the cluster and datum models in this problem.

In visualization, we denote a point $\vec{x}_n$ as a circle with radius linearly relative to ω_n. Here, we assume that all datums belong to $[-X, X]^2$, for a known positive X, to be addressed as the *range* variable. In Figure 4.4, and the other figures in this section, range is denoted by the shaded thick square surrounding the points.

As denoted in Figure 4.4, in this problem class, the two functions $\phi\left(\cdot\right)$ and $\Psi\left(\cdot\right)$ are the Euclidean and the center of mass operators and the set of datums is generated based on the superposition of multiple Gaussian distributions. Moreover, in order to produce the input set of datums, we we first randomly place 8 points in $[-X, X]^2$. Here, $X = 10$. Then, for each point, which we address as $[x_n^\circ, y_n^\circ]$, we generate 200 datums, where we generate the x and y coordinates based on independent $N(x_n^\circ, 1)$ and $N(y_n^\circ, 1)$ distributions. Subsequently, we add 3×200 background points which we generate using two independent $N(0, X)$ distributions. Finally, we reject points outside $[-X, X]^2$. This problem class is defined as non-inclusive and utilizes the scale of 4 and we produce the initial

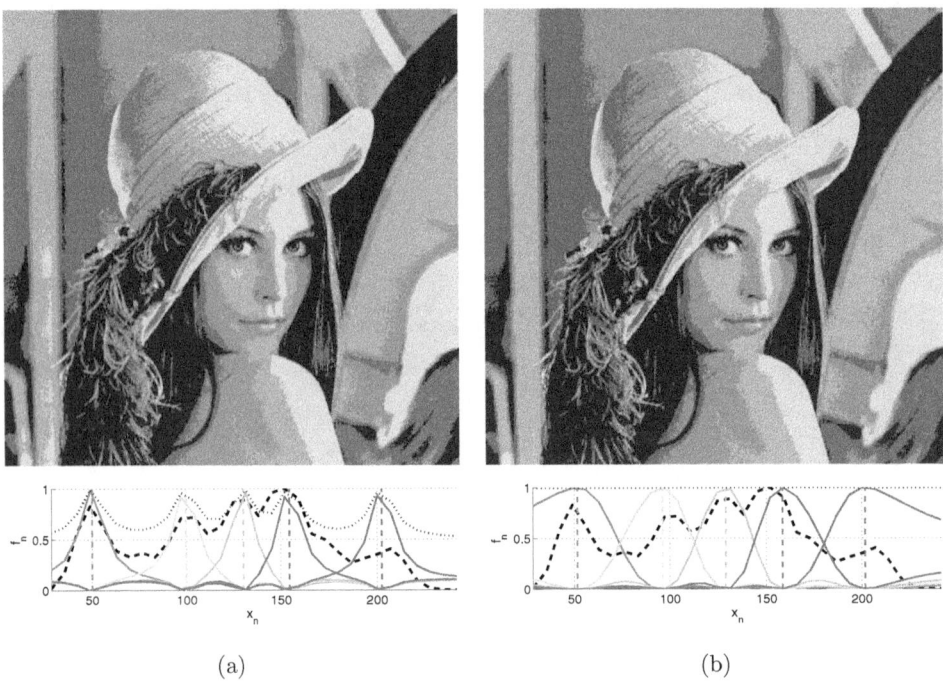

Figure 4.3: Results of the execution of the developed method for grayscale image multi-level thresholding. (a) Developed algorithm. (b) Classical FCM.

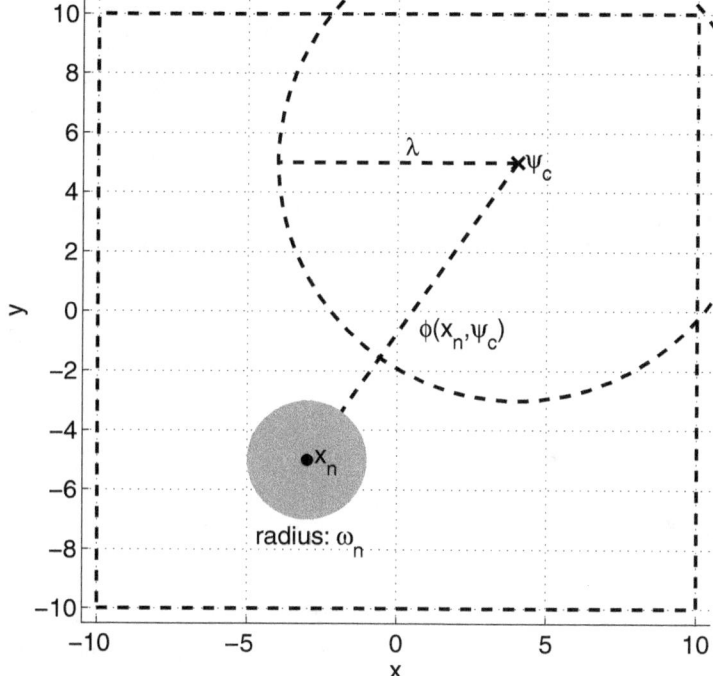

Figure 4.4: Cluster and datum models in the 2-D Euclidean clustering problem class. Here, a cluster is a modeled as a circle in the 2-D space, centered at ψ_c, and datums belong to a square in $\mathbb{R}^2$. Each datum has a corresponding non-negative weight associated to it. The scale, λ, governs the radius of the clusters.

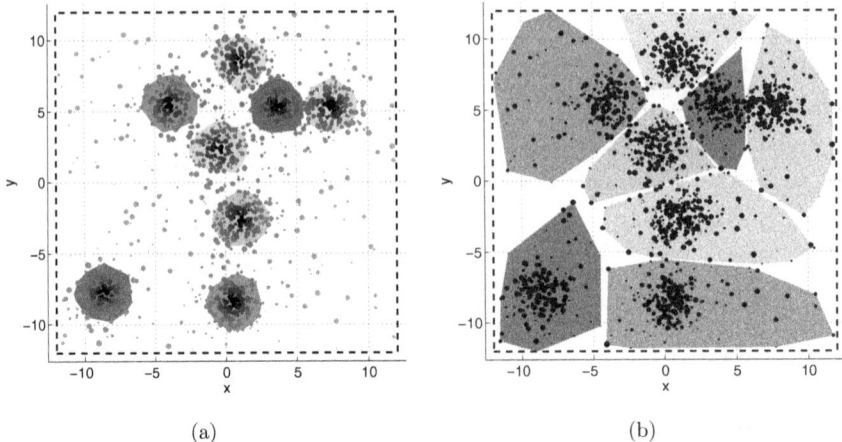

Figure 4.5: Results of the execution of the developed method for 2-D Euclidean clustering. Size of the datums denote their weight and their shade of gray indicates the probability of being an inlier. (a) Developed algorithm. (b) Classical FCM.

set of clusters as $C = 8$ uniformly-distributed points in $[-X, X]^2$. In this problem class we set $\lambda = 4^2$.

Figure 4.5 shows the results of executing the developed algorithm as well as those of the classical FCM on a set of datums. Here, the colored polygons exhibit the convex hull of the datums in each cluster and shades of gray indicate the probability that the corresponding datum is an inlier. Radius of a datum conveys its weight.

We note that while the developed algorithm produces compact clusters, the classical FCM extends the clusters into the entire set of datums. This is a direct result of the outlier assessment process organically embedded in the developed algorithm. Nevertheless, we emphasize, that the outlier rejection mechanism present in the developed algorithm is an emergent property of the utilized clustering model. This, we argue, is a major advantage over the approaches in the literature in which a mechanism involving confidence-thresholding is utilized in order to detect outliers. Such methods require the "prudent" [110, page 452, top of column two] setting of a configuration parameter or the approach will fail to perform outlier rejection efficiently. This phenomenon is also visible in the different shades of gray given to the different datums in the results of the developed algorithm. As seen in Figure 4.5–(a), darker points are considered more probable to be inliers by

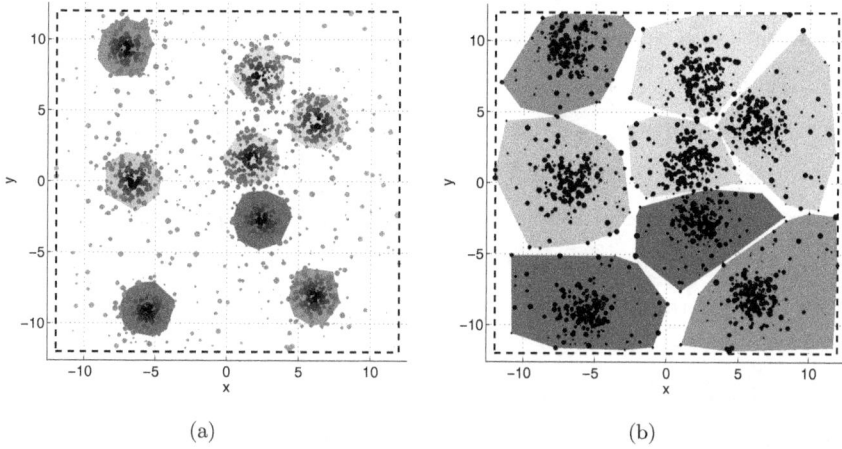

Figure 4.6: Results of the execution of the developed method for 2-D Euclidean clustering. Size of the datums denote their weight and their shade of gray indicates the probability of being an inlier. (a) Developed algorithm. (b) Classical FCM.

the developed algorithm. Hence, we observe a more efficient detection of the boundaries of the clusters and a more optimal separation of the datums into inliers and outliers by the developed algorithm, compared to the classical FCM.

The developed algorithm converges in this case after 19 iterations and 189 milliseconds of operations while the classical FCM requires 31 iterations and 190 milliseconds. These values are not representative however, because experimentation with 100 different sets of datums indicates that the developed algorithm requires 215 ± 81 milliseconds of operation while the classical FCM concludes in 124 ± 56 milliseconds. Hence, in average, the developed algorithm inflates the execution time by a factor of 2.03, compared to the classical FCM. Figure 4.6 shows another set of datums processed by the developed algorithm as well as by FCM.

4.4 Plane Finding in Range Data

In this section, we discuss the problem class which is concerned with finding planar sections in range data. Figure 4.7 provides a visual representation of cluster and datum models in this problem class.

The input set of datums in this problem class contains 3-D points captured by a *Kinect 2* depth acquisition device. The depth-maps used in this experiment are captured at the resolution of 512 × 424 pixels. Note that, the intrinsic parameters of the camera are acquired through the Kinect SDK and that each datum in this problem class has the weight of one.

Clusters in this problem class are defined as planes which have a thickness and the process of fitting a plane to a weighted set of 3-D points is carried out through a weighted variant of Singular Value Decomposition (SVD). This problem class utilizes the scale of 200 millimeters and is defined as non-inclusive, because we expect outliers and other datums which do not belong to any of the planes to exist in the set of datums and wish to identify and exclude them. We initialize the process by placing three planes in the extremities of the room, where based on *a priori* knowledge we expect planes to exist.

Figure 4.8 shows the results of executing the developed algorithm as well as the classical FCM on range data captured from a room with visible floor and walls and three human figures present in it. Here, the pointcloud is viewed from the top while the camera is located at the bottom center of the view.

In Figures 4.8-(a) and (b), points are painted according to the cluster they belong to. We observe in Figures 4.8-(a) that the floor and the two walls are successfully detected by the developed algorithm. Additionally, points belonging to the people present in the scene are discarded by the developed algorithm as outliers, and are hence painted as black. Close assessment of Figure 4.8-(b), however, reveals that the human figures are considered as parts of the different planes detected by the classical FCM. The fact is that FCM has completely failed in detecting the planes present in the scene and has converged to arbitrary locations in the data, except for the floor plane which is partly detected. Nevertheless, this plane is also mixed with portions of one of the walls as well. On the contrary, Figure 4.8-(a) indicates that the developed algorithm robustly detects and locks into the two walls in the scene and the floor. A similar observation is made in other sets of datums, which belong to the same scene captured from different vantage points, as shown in Figure 4.9. As seen here, in every case, the developed method correctly identifies the planes present in the scene while the classical FCM always fails.

For the results shown in Figure 4.8, the developed algorithm converges after 52 iterations and 4228 milliseconds of operations while classical FCM requires 60 iterations and 4329 milliseconds before it concludes. We subsequently utilized a set of 7 samples taken at the same room, where the camera was placed in different positions and the human subjects moved around. In average, the developed algorithm required 2900±819 milliseconds while the classical FCM converged after 2171±

33

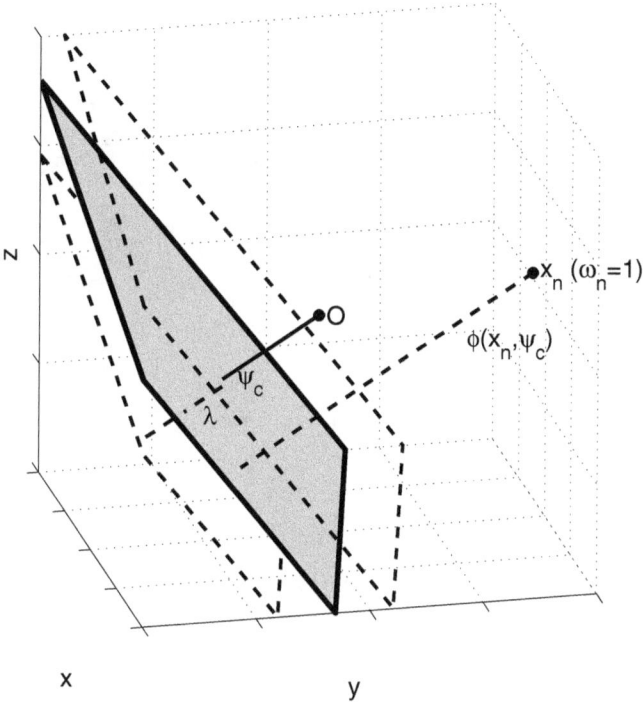

Figure 4.7: Cluster and datum models in the plane finding in range data problem class. Here, a cluster is a modeled as a thick planar section of the 3-D space and is denoted by the closest point on it to the origin (more precise definition given in the text). Datums in this problem class are points in the 3-D space and distance between a datum and a cluster is defined as the distance between the corresponding point and the respective plane. The scale, λ, governs the thickness of the clusters.

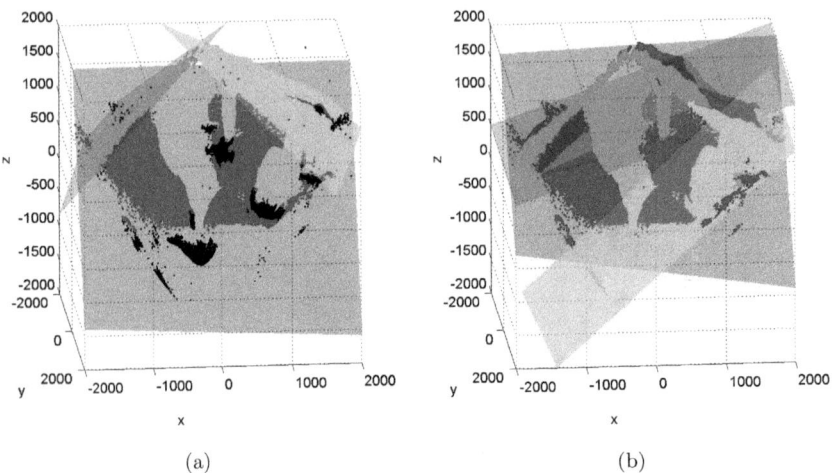

Figure 4.8: Results of the execution of the developed method for plane finding in range data. (a) Developed algorithm. (b) Classical FCM.

1372 milliseconds. Hence, in average, the computational complexity of the developed algorithm is 1.72 times more than that of the classical FCM. This numerical assessment should be reviewed while the failure of FCM in the experiments is also taken into consideration.

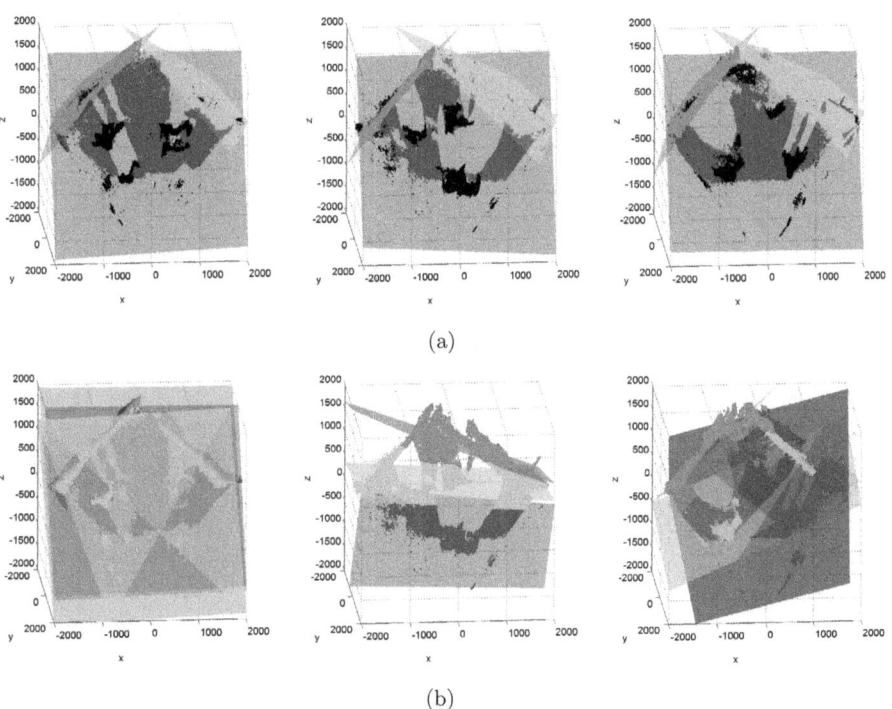

Figure 4.9: Results of the execution of the developed method for plane finding in range data. (a) Developed algorithm. (b) Classical FCM.

Chapter 5

Conclusions

In the past few decades, many approaches to solving the fuzzy/possibilistic clustering problem have been proposed. These different methods utilize different sets of identifiers in order to model concepts such as membership to a cluster, probability of being an inlier, and others. We argue, however, that the literature is dominated by approaches which utilize parallel clustering models. In essence, in these models, a group of clusters process a datum independently, and the results are then passed through a consolidation stage which satisfies constraints such as sum-of-one for membership values. We also argue that works in the literature are generally heavily based on heuristically engineered objective functions and constraints which are intuitively translated from a verbal description of what the clustering algorithm is expected to carry out.

In this work, we develop a novel sequential clustering model. In this model, clusters receive the datums in a sequence and either accept it or pass it on to the next stage. We outline a Bayesian inference framework which we use in order to assess the loss in the model. The assessment carried out in this document is independent of the datum and cluster models of any particular problem class.

We observe that the notion of the probability of being an inlier emerges within the developed clustering model. Hence, in this work, a generic class-independent clustering algorithm is developed which also carries out outlier rejection, if required, with no dependence on any confidence threshold or similar measures.

We observe that, conventionally, suggestions of novel clustering approaches attempt to attain validation through providing experimental results which indicate that the developed technique in fact yields acceptable results. We argue that this after-the-fact approach, while necessary, is based on an epistemological challenge. In this argument, we suggest that the translation of verbal expectations from the clustering problem into formal notations must follow a clear mathematical model. As such, while providing experimental results in this context is a necessity, we argue that such results, alone, do not validate a method in the absence of a sound derivation procedure.

Nevertheless, we considered three problem classes and outlined the utilization of the developed method for them. In fact, for each problem class, we presented a number of problem instances and compared the results of the developed algorithm on these samples to what is produced by the classical FCM. In this work, the developed algorithm and the classical FCM were executed in completely identical circumstances.

We observed that the classical FCM extends the clusters in order to cover the entire set of datums, whereas the developed method utilizes its emerged outlier rejection mechanism in order to detect and exclude outliers. We also showed that the robust loss function utilized in the developed algorithm allows it to focus on dense clusters and to avoid adjacent distractions. These benefits are achieved by the developed algorithm while its execution time is less than four times that of the classical FCM. In fact, in certain cases, the developed algorithm functions faster than the classical FCM, mainly because convergence is achieved in a fewer number of iterations by the developed algorithm compared to the classical FCM.

Acknowledgments

The author wishes to thank the management of Epson Edge for their help and support during the course of this research. We wish to thank Mahsa Pezeshki for proofreading this manuscript. The idea for this research was conceived at *The 3 Brewers* pub in Toronto while enjoying the night with a group of friends. A view of the napkin which contained the first version of the mathematics of this work is carried at the beginning of this document. We wish to thank the respective anonymous reviewers of this document for their constructive comments and recommendations. We wish to thank friends on Twitter, Facebook, and Google+ for their help in finding key references for this document.

Bibliography

[1] A. Abadpour, A sequential bayesian alternative to the classical parallel fuzzy clustering model, Information Sciences 318 (2015) 28–47.

[2] J. B. MacQueen, Some methods for classification and analysis of multivariate observations, in: Proceedings of 5-th Berkeley Symposium on Mathematical Statistics and Probability, Berkeley, 1967, pp. 281–297.

[3] R. Gray, Y. Linde, Vector quantizers and predictive quantizers for Gauss-Markov sources, IEEE Transactions on Communications 30 (2) (1982) 381–389.

[4] T. Kanungo, D. M. Mount, N. S. Netanyahu, C. D. Piatko, R. Silverman, A. Y. Wu, An efficient k-means clustering algorithm: Analysis and implementation, IEEE Transactions on Pattern Analysis and Machine Intelligence 24 (7) (2002) 881–892.

[5] G. H. Ball, D. J. Hall, A clustering technique for summarizing multivariate data, Behavioral Science 12 (2) (1967) 153–155.

[6] L. A. Zadeh, Fuzzy sets, Information Control 8 (1965) 338–353.

[7] A. Baraldi, P. Blonda, A survey of fuzzy clustering algorithms for pattern recognition. I & II, IEEE Transactions on Systems, Man, and Cybernetics, Part B: Cybernetics 29 (6) (1999) 778–801.

[8] R. Duda, P. Hart, Pattern Classification and Scene Analysis, Wiley, New York, 1973.

[9] J. C. Bezdek, Pattern Recognition with Fuzzy Objective Function Algorithms, Plenum Press, New York, 1981.

[10] E. H. Ruspini, A new approach to clustering, Information & Control 15 (1) (1969) 22–32.

[11] J. C. Dunn, A fuzzy relative of the ISODATA process and its use in detecting compact well-separated clusters, Journal of Cybernetics 3 (3) (1973) 32–57.

[12] J. M. Leski, Generalized weighted conditional fuzzy clustering, IEEE Transactions on Fuzzy Systems 11 (6) (2003) 709–715.

[13] J. Yu, Q. Cheng, H. Huang, Analysis of the weighting exponent in the FCM, IEEE Transactions on Systems, Man, and Cybernetics, Part B: Cybernetics 34 (1) (2004) 634–639.

[14] M. Trivedi, J. C. Bezdek, Low-level segmentation of aerial images with fuzzy clustering, IEEE Transactions on Systems, Man, and Cybernetics 16 (4) (1986) 589–598.

[15] H. Frigui, R. Krishnapuram, A robust algorithm for automatic extraction of an unknown number of clusters from noisy data, Pattern Recognition Letters 17 (12) (1996) 1223–1232.

[16] F. Klawonn, R. Kruse, H. Timm, Fuzzy shell cluster analysis, in: G. della Riccia, H. Lenz, R. Kruse (Eds.), Learning, networks and statistics, Springer, 1997, pp. 105–120.

[17] J. C. Bezdek, A physical interpretation of fuzzy ISODATA, IEEE Transactions on Systems, Man and Cybernetics SMC-6 (5) (1976) 387–389.

[18] N. R. Pal, J. C. Bezdek, On cluster validity for the fuzzy C-means model, IEEE Transactions on Fuzzy Systems 3 (3) (1995) 370–379.

[19] K. Zhou, C. Fu, S. L. Yang, Fuzziness parameter selection in fuzzy c-means: The perspective of cluster validation, Science China Information Sciences 57 (11) (2014) 1–8.

[20] M. Lichman, UCI machine learning repository (2013).
URL http://archive.ics.uci.edu/ml

[21] J. C. Bezdek, N. R. Pal, Some new indexes of cluster validity, IEEE Transactions on Systems, Man, and Cybernetics, Part B: Cybernetics 28 (3) (1998) 301–315.

[22] I. Sledge, J. C. Bezdek, T. C. Havens, J. M. Keller, Relational generalizations of cluster validity indices, IEEE Transactions on Fuzzy Systems 18 (4) (2010) 771–786.

[23] I. Ozkan, I. Turksen, Upper and lower values for the level of fuzziness in FCM, in: P. P. Wang, D. Ruan, E. E. Kerre (Eds.), Fuzzy Logic, Vol. 215 of Studies in Fuzziness and Soft Computing, Springer Berlin Heidelberg, 2007, pp. 99–112.

[24] K.-L. Wu, Analysis of parameter selections for fuzzy c-means, Pattern Recognition 45 (1) (2012) 407–415.

[25] P. J. Rousseeuw, E. Trauwaert, L. Kaufman, Fuzzy clustering with high contrast, Journal of Computational and Applied Mathematics 64 (1-2) (1995) 81–90.

[26] F. Klawonn, F. Hoppner, What is fuzzy about fuzzy clustering? Understanding and improving the concept of the fuzzifier, in: M. R. Berthold, H.-J. Lenz, E. Bradley, R. Kruse, C. Borgelt (Eds.), Advances in Intelligent Data Analysis V, Vol. 2810 of Lecture Notes in Computer Science, Springer Berlin Heidelberg, 2003, pp. 254–264.

[27] F. Klawonn, Fuzzy clustering: Insights and a new approach, Mathware and soft computing 11 (2004) 125–142.

[28] W. Pedrycz, Conditional fuzzy C-means, Pattern Recognition Letters 17 (6) (1996) 625–631.

[29] W. Pedrycz, Fuzzy set technology in knowledge discovery, Fuzzy Sets and Systems 98 (3) (1998) 279–290.

[30] W. Pedrycz, Conditional fuzzy clustering in the design of radial basis function neural networks, IEEE Transactions on Neural Networks 9 (4) (1998) 601–612.

[31] K. K. Chintalapudi, M. Kam, The credibilistic fuzzy C-means clustering algorithm, in: IEEE International Conference on Systems, Man, and Cybernetics (SMC 1998), Vol. 2, 1998, pp. 2034–2039.

[32] M.-S. Yang, K.-L. Wu, Unsupervised possibilistic clustering, Pattern Recognition 39 (1) (2006) 5–21.

[33] J. Noordam, W. van den Broek, L. Buydens, Multivariate image segmentation with cluster size insensitive fuzzy C-means, Chemometrics and Intelligent Laboratory Systems 64 (1) (2002) 65–78.

[34] R. Kruse, C. Doring, M.-J. Lesot, Fundamentals of fuzzy clustering, in: J. V. de Oliveira, W. Pedrycz (Eds.), Advances in Fuzzy Clustering and its Applications, Wiley, England, 2007, pp. 3–29.

[35] R. Yager, D. Filev, Approximate clustering via the mountain method, IEEE Transactions on Systems, Man and Cybernetics 24 (8) (1994) 1279–1284.

[36] G. Beni, X. Liu, A least biased fuzzy clustering method, IEEE Transactions on Pattern Analysis and Machine Intelligence 16 (9) (1994) 954–960.

[37] K. Rose, E. Gurewitz, G. Fox, Constrained clustering as an optimization method, IEEE Transactions on Pattern Analysis and Machine Intelligence 15 (8) (1993) 785–794.

[38] J. M. Leski, Fuzzy c-varieties/elliptotypes clustering in reproducing kernel Hilbert space, Fuzzy Sets and Systems 141 (2) (2004) 259–280.

[39] D.-M. Tsai, C.-C. Lin, Fuzzy C-means based clustering for linearly and nonlinearly separable data, Pattern Recognition 44 (8) (2011) 1750–1760.

[40] K.-L. Wu, M.-S. Yang, Alternative C-means clustering algorithms, Pattern Recognition 35 (10) (2002) 2267–2278.

[41] L. Chen, C. Chen, M. Lu, A multiple-kernel fuzzy C-means algorithm for image segmentation, IEEE Transactions on Systems, Man, and Cybernetics, Part B: Cybernetics 41 (5) (2011) 1263–1274.

[42] S. Chen, D. Zhang, Robust image segmentation using FCM with spatial constraints based on new kernel-induced distance measure, IEEE Transactions on Systems, Man, and Cybernetics, Part B: Cybernetics 34 (4) (2004) 1907–1916.

[43] K. Honda, N. Sugiura, H. Ichihashi, Fuzzy PCA-guided robust k-means clustering, IEEE Transactions on Fuzzy Systems 18 (1) (2010) 67–79.

[44] H. Zha, C. Ding, M. Gu, X. He, H. Simon, Spectral relaxation for K-means clustering, in: Proceedings of Advances in Neural Information Processing Systems, 2002, pp. 1057–1064.

[45] R. J. Hathaway, J. W. Davenport, J. C. Bezdek, Relational duals of the C-means clustering algorithms, Pattern Recognition 22 (2) (1989) 205–212.

[46] R. J. Hathaway, J. C. Bezdek, NERF C-means: Non-Euclidean relational fuzzy clustering, Pattern Recognition 27 (3) (1994) 429–437.

[47] S. Nascimento, B. Mirkin, F. Moura-Pires, Multiple prototype model for fuzzy clustering, in: D. J. Hand, J. N. Kok, M. R. Berthold (Eds.), Advances in Intelligent Data Analysis, Vol. 1642 of Lecture Notes in Computer Science, Springer Berlin Heidelberg, 1999, pp. 269–279.

[48] K. Jajuga, L_1-norm based fuzzy clustering, Fuzzy Sets and Systems 39 (1) (1991) 43–50.

[49] L. Bobrowski, J. C. Bezdek, C-means clustering with the ℓ_1 and ℓ_∞ norms, IEEE Transactions on Systems, Man, and Cybernetics 21 (3) (1991) 545–554.

[50] R. J. Hathaway, J. C. Bezdek, Optimization of clustering criteria by reformulation, IEEE Transactions on Fuzzy Systems 3 (1995) 241–246.

[51] R. J. Hathaway, J. C. Bezdek, Y. Hu, Generalized fuzzy C-means clustering strategies using L_p norm distances, IEEE Transactions on Fuzzy Systems 8 (5) (2000) 576–582.

[52] N. B. Karayiannisa, M. M. Randolph-Gips, Non-Euclidean C-means clustering algorithms, Intelligent Data Analysis 7 (2003) 405–425.

[53] D. E. Gustafson, W. C. Kessel, Fuzzy clustering with a fuzzy covariance matrix, in: IEEE Conference on Decision and Control including the 17th Symposium on Adaptive Processes, Vol. 17, San Diego, CA, 1979, pp. 761–766.

[54] I. Gath, A. Geva, Unsupervised optimal fuzzy clustering, IEEE Transaction on Pattern Analysis Machine Intelligence 11 (7) (1989) 773–781.

[55] H. Frigui, R. Krishnapuram, A comparison of fuzzy shell-clustering methods for the detection of ellipses, IEEE Transactions on Fuzzy Systems 4 (2) (1996) 193–199.

[56] R. Krishnapuram, H. Frigui, O. Nasraoui, Fuzzy and possibilistic shell clustering algorithms and their application to boundary detection and surface approximation - Parts I & II, IEEE Transaction on Fuzzy Systems 3 (1) (1995) 29–60.

[57] R. N. Dave, R. Krishnapuram, Robust clustering methods: A unified view, IEEE Transactions on Fuzzy Systems 5 (2) (1997) 270–293.

[58] J. Leski, Towards a robust fuzzy clustering, Fuzzy Sets and Systems 137 (2) (2003) 215–233.

[59] P. D'Urso, L. D. Giovanni, Robust clustering of imprecise data, Chemometrics and Intelligent Laboratory Systems 136 (2014) 58–80.

[60] J. J. D. Gruijter, A. B. McBratney, A modified fuzzy K-means method for predictive classification, in: H. H. Bock (Ed.), Classification and Related Methods of Data Analysis, Elsevier, Amsterdam, The Netherlands, 1988, pp. 97–104.

[61] R. N. Dave, Characterization and detection of noise in clustering, Pattern Recognition Letters 12 (11) (1991) 657–664.

[62] Y. Ohashi, Fuzzy clustering and robust estimation, Presented at the 9th SAS Users Group International (SUGI) Meeting at Hollywood Beach, Florida. (1984).

[63] R. N. Dave, Robust fuzzy clustering algorithms, in: Second IEEE International Conference on Fuzzy Systems, Vol. 2, 1993, pp. 1281–1286.

[64] J. M. Jolion, P. Meer, S. Bataouche, Robust clustering with applications in computer vision, IEEE Transactions on Pattern Analysis and Machine Intelligence 13 (8) (1991) 791–802.

[65] S. Zhuang, T. Wang, P. Zhang, A highly robust estimator through partially likelihood function modeling and its application in computer vision, IEEE Transactions on Pattern Analysis and Machine Intelligence 14 (1) (1992) 19–35.

[66] R. Krishnapuram, J. M. Keller, A possibilistic approach to clustering, IEEE Transactions on Fuzzy Systems 1 (2) (1993) 98–110.

[67] M. Barni, V. Cappellini, A. Mecocci, Comments on "A possibilistic approach to clustering", IEEE Transactions on Fuzzy Systems 4 (3) (1996) 393–396.

[68] H. Timm, C. Borgelt, C. Doring, R. Kruse, An extension to possibilistic fuzzy cluster analysis, Fuzzy Sets and Systems 147 (1) (2004) 3–16.

[69] R. Dave, S. Sen, On generalising the noise clustering algorithms, in: Proceedings of the 7th IFSA World Congress (IFSA 1997), 1997, pp. 205–210.

[70] N. R. Pal, K. Pal, J. C. Bezdek, A mixed c-means clustering model, in: Proceedings of the Sixth IEEE International Conference on Fuzzy Systems, Vol. 1, 1997, pp. 11–21.

[71] N. R. Pal, K. Pal, J. M. Keller, J. C. Bezdek, A new hybrid C-means clustering model, in: Proceedings of the 2004 IEEE International Conference on Fuzzy Systems, Vol. 1, 2004, pp. 179–184.

[72] X.-Y. Wang, J. M. Garibaldi, Simulated annealing fuzzy clustering in cancer diagnosis, Informatica 29 (1) (2005) 61–70.

[73] A. Keller, Fuzzy clustering with outliers, in: Proceesings of the 19th International Conference of the North American Fuzzy Information Processing Society (NAFIPS 2000), 2000, pp. 143–147.

[74] D.-Q. Zhang, S.-C. Chen, A comment on "Alternative C-means clustering algorithms", Pattern Recognition 37 (2) (2004) 173–174.

[75] A. K. Jain, R. C. Dubes, Algorithms for Clustering Data, Prentice-Hall, 1981.

[76] R. Krishnapuram, C.-P. Freg, Fitting an unknown number of lines and planes to image data through compatible cluster merging, Pattern Recognition 25 (4) (1992) 385–400.

[77] R. Krishnapuram, H. Frigui, O. Nasraoui, Quadric shell clustering algorithms and their applications, Pattern Recognition Letters 14 (7) (1993) 545–552.

[78] R. Krishnapuram, O. Nasraoui, H. Frigui, The fuzzy C-spherical shells algorithm: a new approach, IEEE Transactions on Neural Networks 3 (5) (1992) 663–671.

[79] R. Krishnapuram, Generation of membership functions via possibilistic clustering, in: IEEE World Congress on Computational Intelligence, 1994, pp. 902–908 vol.2.

[80] R. N. Dave, T. Fu, Robust shape detection using fuzzy clustering: Practical applications, Fuzzy Sets and Systems 65 (2-3) (1994) 161–185.

[81] C. Stewart, MINPRAN: A new robust estimator for computer vision, IEEE Transactions on Pattern Analysis and Machine Intelligence 17 (10) (1995) 925–938.

[82] H. Frigui, R. Krishnapuram, Clustering by competitive agglomeration, Pattern Recognition 30 (7) (1997) 1109–1119.

[83] P. D'Urso, Fuzzy clustering of fuzzy data, in: J. V. de Oliveira, W. Pedrycz (Eds.), Advances in Fuzzy Clustering and its Applications, Wiley, England, 2007, pp. 155–192.

[84] A. Abadpour, A. S. Alfa, J. Diamond, Video-on-demand network design and maintenance using fuzzy optimization, IEEE Transactions on Systems, Man, and Cybernetics, Part B: Cybernetics 38 (2) (2008) 404–420.

[85] L. Szilagyi, Z. Benyo, S. Szilagyi, H. S. Adam, MR brain image segmentation using an enhanced fuzzy C-means algorithm, in: Proceedings of the 25th Annual International Con-

ference of the IEEE Engineering in Medicine and Biology Society (EMBS 2003), Vol. 1, 2003, pp. 724–726.

[86] R. J. Hathaway, Y. Hu, Density-weighted fuzzy C-means clustering, IEEE Transactions on Fuzzy Systems 17 (1) (2009) 243–252.

[87] Y. Yang, Image segmentation based on fuzzy clustering with neighborhood information, Optica Applicata 39 (1) (2009) 135–147.

[88] R. Nock, F. Nielsen, On weighting clustering, IEEE Transactions on Pattern Analysis and Machine Intelligence 28 (8) (2006) 1223–1235.

[89] C.-H. Li, W.-C. Huang, B.-C. Kuo, C.-C. Hung, A novel fuzzy weighted C-means method for image classification, International Journal of Fuzzy Systems 10 (3) (2008) 168–173.

[90] C.-C. Hung, S. Kulkarni, B.-C. Kuo, A new weighted fuzzy C-means clustering algorithm for remotely sensed image classification, IEEE Journal of Selected Topics in Signal Processing 5 (3) (2011) 543–553.

[91] K. Tsuda, M. Minoh, K. Ikeda, Extracting straight lines by sequential fuzzy clustering, Pattern Recognition Letters 17 (6) (1996) 643–649.

[92] F. L. Chung, T. Lee, Fuzzy competitive learning, Neural Networks 7 (3) (1994) 539–551.

[93] G. L. Zheng, An enhanced sequential fuzzy clustering algorithm, International Journal of Systems Science 30 (3) (1999) 295–307.

[94] P. W. Holland, R. E. Welsch, Robust regression using iteratively reweighted least squares, Communication Statistics - Theory and Methods A6 (9) (1977) 813–827.

[95] R. Dutter, Numerical solution of robust regression problems: Ccomputational aspects, a comparison, Journal of Statistical Computation and Simulation 5 (3) (1977) 207–238.

[96] P. J. Huber, E. Ronchetti, Robust Statistics, Wiley, New York, 2009.

[97] A. E. Beaton, J. W. Tukey, The fitting of power series, meaning polynomials, illustrated on band-spectroscopic data, Technometrics 16 (1974) 147–185.

[98] F. R. Hampel, E. M. Ponchotti, P. J. Rousseeuw, W. A. Stahel, Robust Statistics: The Approach based on Influence Functions, Wiley, New York, 2005.

[99] J. M. Leski, Fuzzy c-ordered-means clustering, Fuzzy Sets and Systems.

[100] E. Weiszfeld, Sur le point pour lequel la somme des distances de n points donnes est minimum, Tohoku Mathematical Journal 43 (1937) 355–386.

[101] W. Miehle, Link-length minimization in networks, Operations Research 6 (2) (1958) 232–243.

[102] J. B. Rosen, G. L. Xue, On the convergence of Miehle's algorithm for the Euclidean multifactory location problem, Operations Research 40 (1) (1992) 188–191.

[103] H. W. Kuhn, R. E. Kuenne, An efficient algorithm for the numerical solution of the generalized Weber problem in the spatial economics, Journal of Regional Science 4 (1962) 21–33.

[104] L. Cooper, Location-allocation problems, Operation Research 11 (1963) 331–343.

[105] Z. Drezner, A note on accelerating the Weiszfeld procedure, Location Science 3 (1995) 275–279.

[106] J. J. More, The Levenberg-Marquardt algorithm: Implementation and theory, in: G. Watson (Ed.), Numerical Analysis, Vol. 630 of Lecture Notes in Mathematics, Springer Berlin Heidelberg, 1978, pp. 105–116.

[107] A. Abadpour, Rederivation of the fuzzypossibilistic clustering objective function through Bayesian inference, Fuzzy Sets and Systems 305 (2016) 29–53.

[108] J. Noordam, W. H. A. M. van den Broek, Multivariate image segmentation based on geometrically guided fuzzy C-means clustering, Journal of Chemometrics 16 (1) (2002) 1–11.

[109] P. Arbelaez, M. Maire, C. Fowlkes, J. Malik, Contour detection and hierarchical image segmentation, IEEE Transactions on Pattern Analysis and Machine Intelligence 33 (5) (2011) 898–916.

[110] H. Frigui, R. Krishnapuram, A robust competitive clustering algorithm with applications in computer vision, IEEE Transactions on Pattern Analysis and Machine Intelligence 21 (5) (1999) 450–465.

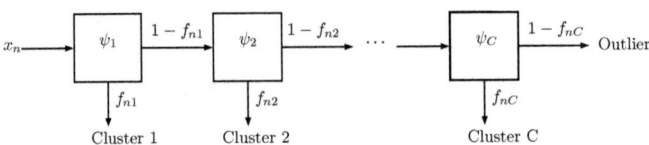

Figure 5.1: The sequence-of-clusters model utilized in the algorithm developed in this work. Here, a sequence of C clusters process an input datum and produce C membership values and a probability value that indicates whether or not the datum is an outlier. Note that probability values are conditional.

www.ingramcontent.com/pod-product-compliance
Lightning Source LLC
Chambersburg PA
CBHW061447180526
45170CB00004B/1597